AMERICAN DEMOCRACY IN JEOPARDY

A Nation of People Vulnerable to Being Told What to Think

FRANK DALOTTO

authorHOUSE®

AuthorHouse™
1663 Liberty Drive
Bloomington, IN 47403
www.authorhouse.com
Phone: 1-800-839-8640

First published by AuthorHouse 5/4/2010

ISBN: 978-1-4490-7760-0 (e)
ISBN: 978-1-4490-7758-7 (sc)
ISBN: 978-1-4490-7759-4 (hc)

Library of Congress Control Number: 2010905471

Printed in the United States of America
Bloomington, Indiana

This book is printed on acid-free paper.

Dedication

Dedicated to America's children and young adults and especially my grandchildren—Alexander, Katie, Emmy, Kristin, Matthew, Tommy, Nicholas, Ophelia, Stella, and Lucy—who will be fortunate to acquire the critical thinking skills necessary to effectively contribute to preserving American democracy in the twenty-first century.

"People used to say, 'Ignorance is no excuse.' Today, ignorance is no problem. Our schools promote so much self-esteem that people confidently spout off about all sorts of things that they know nothing about."
  —Thomas Sowell, PhD, "Random Thoughts," August 12, 2004

"To the intelligent man or woman, life appears infinitely mysterious. But the stupid have an answer for every question."

  —Edward Abbey

CONTENTS

PROLOGUE

"Never make the mistake of thinking that you know everything about anything" —Chinese proverb

I grew up in the 1940s and early 1950s in the South Beach section of Staten Island, New York, a close-knit community made up largely of first- and second-generation Italians, who were similar to other large immigrant communities in that they hung on to their ethnic roots and culture and were very slow in assimilating to the "American way of life."

We lived in a multi-family apartment on several acres of land owned by my grandparents, who occupied one of the apartments. My uncle and aunt occupied the other. This was not unusual at the time, as many Italian immigrant families chose to remain in close contact with each other.

My grandparents spoke very little English and would carry on conversations entirely in Italian with family members and friends. They strictly observed Italian traditions of large Sunday and holiday family gatherings, with daylong tastings of food and wine. Tradition at that time dictated that the women prepared, cooked, and served the meal and cleaned up afterward, while the men enjoyed their day of rest discussing work, politics, and how their kids and grandkids were drifting away from family traditions. They did this while drinking homemade wine, smoking smelly Italian cigars, and in the summer, playing bocce.

My first realization that not all people are of the same culture and beliefs came when I went to high school outside the community and found that not everyone is Italian; that not everyone lives in a home that has a statue of the Virgin Mary, fig trees, tomato plants, and a bocce court; and that not everyone goes to Catholic church every Sunday.

Although I took on new non-Italian friends, the influence from family and South Beach friends was strong, and the values and thinking patterns remained intact.

As I grew into adulthood, I became more exposed to other cultures, lifestyles, and beliefs. I began to wonder why people living in the same city could have beliefs that were so different from each other. Was it right or wrong, or just differences?

My next exposure to different cultures and beliefs came when I went to the University of Missouri and began to associate with students having Midwestern, non-Italian backgrounds. They had set beliefs about Italians that were taken from TV, the movies, family, and friends. It was a constant struggle to convince them that not all Italians belong to the Mafia, have greasy hair, and eat nothing but spaghetti and pizza.

My college years at the University of Missouri occurred during the Vietnam War. Student antiwar demonstrations created turbulent times on campus. My family beliefs and roots instilled in me the belief that it was un-American and unpatriotic to question the government and criticize the war. With day-to-day exposure to a different point of view, I slowly began to see the merit in other points of view.

Over time, I continued to associate with more people having strongly opposed beliefs.

In the early 1970s, when I was enrolled in a masters degree program and had to complete a thesis, I gained a knowledge and appreciation of the need to seek out firsthand, qualified sources of research data. I also learned that there are techniques such as regression analysis and other ways to evaluate data to ensure that modeling, identification, and analysis of variables are taken into account before conclusions are drawn.

In the mid-seventies, I was employed by ITT. After receiving my master's degree, I was placed on a track for executive management grooming and promoted to engineering manager. I was selected to represent ITT and spend two months in Japan in a management employee exchange program with Japan's international communications company (KDD).

ITT was in the business of providing international cable and satellite networks to corporations and the U.S. Department of Defense. My engineering staff was responsible for designing and installing the U.S. end of the network, and we had to work closely with an overseas international communications partner to install the overseas end.

In order to work cooperatively with our overseas partners to complete the network design projects, we needed to have an understanding and appreciation of each other's culture, beliefs, and business practices.

During my two-month period in Japan, I was fortunate to visit virtually every corner of Japan. This was accomplished with three-day trips to each of Japan's important commerce centers and remote villages where they have communication centers located. On each of these trips I was assigned a Japanese management employee who served as my guide and counterpart. We got to spend many days and evenings together getting to know each other and exchanging ideas.

The most memorable times were spent during the evenings with Japanese management employees who entertained in groups of six to ten at their favorite geisha house, where we got to know each other in a relaxed setting of food, sake, and whiskey.

Being exposed to different beliefs in a different country led me to understand that the reporting and documenting of an historic event depends who writes the story. Near the end of my assignment in Japan, I was asked if there was an area or activity in Japan that we had not visited or covered that I would like to go to.

This was at a period of time that the Second World War was still fresh in everyone's minds, including mine and my Japanese counterparts. We all had some living family members and business associates that had fought in the war.

The bombing of the industrial city of Hiroshima was key to the ending of the war and was widely viewed in this country as a proud American achievement. It stood as an example of America's technological supremacy put to good use. Rightly so, for it ended a terrible world war and ultimately saved the lives of many American soldiers.

I was very interested in seeing Hiroshima firsthand and was puzzled as to why this historic city, the scene of a world-changing event, was not included on the agenda. I therefore, requested we take several days to visit the Hiroshima communications facility, where I could witness the place firsthand and get a sense of the impact of the atomic bombing on the Japanese.

I was met with stunned silence, as they could not comment on or honor my request until they presented it and received the okay from senior management.

After several days of deliberation my request was honored, but they carefully explained what the consequences might be. Hiroshima had previously been on the agenda of the exchange program. However, a previous American ITT exchange manager had a horrible experience with the visit. He angrily accused the Japanese of embarrassing him with the visit to Hiroshima. As a result, he asked to leave the program and return home.

Hiroshima was the scene of horrific devastation to a major Japanese city. One nuclear bomb had killed up to 200,000 innocent people and left a major industrial city completely in ruins. It was only a few years before I arrived that the city was cleared for rebuilding and reoccupation after the long-term effects of radiation contamination subsided to safe levels.

The museum in Hiroshima does an excellent job in documenting the effects of the atomic bomb with artifacts, documents, and pictures, while at the same time downplaying the political aspect and the source of the bomb.

As Americans, however, we were limited to secondhand accounts by newspapers and magazines of the impact.

I found the visit to Hiroshima to be an unforgettable experience and was fortunate to get a firsthand glimpse of the devastation. It was an eye-opener or more appropriately a mind-opener that exposed me to another side of the story. It was key in my learning that it is important we not shut down our opportunities for exploring and learning because of preexisting beliefs. The bombing of Hiroshima ranks as one of the most devastating manmade disasters ever inflicted on innocent people.

The museum in Hiroshima did an excellent recreation of the devastation from actual artifacts, photos, and images. In visiting the museum and in touring the city, I was able to witness the huge death toll and horrific pain and suffering of innocent citizens in a city of 300,000. While the real mortality of the atomic bomb will never be known, estimates from many sources, including the United Nations, the city of Hiroshima, and others, report that 65,000 to 80,000 people died from the initial heat and radiation effects and 69 percent of the city's buildings were destroyed.

Worse yet, there were countless other people who suffered long-term health problems associated with radiation exposure, such as lingering psychological effects and the onset of cancer. In just five years, the death toll was estimated to have reached 150,000 to 200,000, as cancer and other long-term effects took hold. There is no telling how you measure the long-term psychological effects on the remaining 100,000 to 150,000 innocent victims of Hiroshima.

I can't speculate as to the reason why the other American manager became angry for being taken to Hiroshima. Was it because he had a personal attachment and experienced feelings of guilt or embarrassment? Or was it because of some other factor?

The point here is that two Americans, close in age, both from the same corporate environment and both on a track to executive management, were exposed to the same information and original artifacts. One of them found it to be an overwhelmingly negative experience, and the other found it to be a memorable positive experience. Why? Could it be because of a difference in preconditioned or strongly held beliefs?

To this day, the memory of my visit to Hiroshima is still fresh in my mind. When other mankind-inflicted tragedies occur, I can't help but make comparisons using Hiroshima as a reference. One that comes to mind is the tragedy of 9/11 (September 11, 2001), where 2,973 people died because nineteen al-Qaeda members hijacked four commercial passenger jet airliners and crashed them into the World Trade Center in New York City, the Pentagon, and a field in Pennsylvania. This tragedy pales in comparison to Hiroshima in the number of casualties and the devastation to the cities.

The significance to me is that public awareness of 9/11, in the United States and the rest of the world far outpaces that of Hiroshima. At the time of this writing, almost nine years have passed since 9/11. Almost nine years after the Hiroshima bombing on August 6, 1945, I was in high school. I recall that public awareness of the Hiroshima bombing had significantly waned to the point where it was rarely discussed in schools, except during a history lesson on World War II. Nor was it mentioned much in the print media. Recognizing that 9/11 occurred on U.S. soil and was inflicted by foreigners and Hiroshima occurred on foreign soil and was inflicted by the United States, it is evident that the emotions and accounting of events have a great deal to do with who writes the history books and on which side of the war you are on. I believe there are other factors that also account for this huge disparity in public awareness.

In 1945, television in U.S. households was rare, and it was almost nonexistent throughout the rest of the world. Accounting of the incident occurred in print by newspapers and magazines, with black and white photos. Movie theaters offered the only coverage of the event in black and white video. You had to leave the comfort of your home and pay to see movie documentary coverage. It was not affordable for many people in the United States, and certainly not affordable in many other parts of the world. There was no telephone coverage of the event from the site of the disaster to the public. While most U.S. households had at least a crude form of telephone service, telephones were virtually nonexistent in the rest of the world.

In 2001, coverage of 9/11 occurred in real-time color video on TV throughout the world. Nearly everyone in the world, and certainly in the United States, has seen a color video of the 9/11 incident. There were also cable TV news political entertainers who found this exactly the kind of sensational, emotional content that they could feed into, enhancing the shock value of the incident. In 2001 there was also the Internet, with Web sites and e-mails that carried reporting of the event, commentary, and opinion throughout the world. Cell phones and widespread landline telephones were also available in the United States and the rest of the world. In addition to all of this, in 2001 and today, there are a number of well-funded special-interest groups maintaining public awareness of and influencing public opinion on 9/11.

There is no doubt in my mind that had the media and communications technology, cable TV news political entertainers, and well-funded special-interest groups been in place in 1945, Hiroshima would have taken a much higher profile in world public awareness and world public opinion.

Later on in my career, while I was head of Global Telecommunications for AIG, the technology division was undergoing dramatic change. There was great resistance to adapting to new technology from many of the older technology employees who wanted to cling to the old way of doing things and to their existing beliefs and paradigms.

To deal with the resistance from the older technology employees, AIG brought in consultants who ran seminars and workshops on how the mind functions, how beliefs are formed, and how we can learn to accept and appreciate the importance of investigating the merits of different viewpoints. We learned, for example, that people become hypnotized throughout their lives by the thoughts, views, and beliefs they accept from others. When a person believes something to be true, regardless of whether or not it is true, they will act as if it is true and look for evidence to support that belief.

This experience gave me the opportunity to learn about why people behave the way they do and how they go about dealing with exposure to new or unfamiliar beliefs and ways of thinking. I came to understand that people only see and hear what they want to.

For example, with the use of psychological or optical illusions, we were able to discover that one person can see something entirely different in the same picture from someone else. We found that different people looking at an image or picture will oftentimes see different things and both will be right, and that it is possible to have different points of view based on different interpretations of things.

The message here is that there are two sides to every story. For one person to be right doesn't mean the other person has to be wrong.

"We learn from history that we do not learn from history." —George Wilhelm Hegel

During my elementary school years, cowboy and Indian movies were very popular. These movies always portrayed the cowboys and the U.S. Cavalry as the good guys and the Indians as the bad guys. At that time the movie studios were owned by Caucasian Americans and their customers were mostly Caucasian Americans. The cowboy and Indian movies were very popular and profitable during this era and fed into what the moviegoers wanted to see and hear. The movie studios perpetuated the stereotype that the Indians were ruthless savages who attacked cowboys, frontier settlements, and government outposts.

The history books being used in the American school system at that time were written by Caucasian Americans and used by Caucasian American schoolteachers, who literally taught what the textbooks contained, thereby reinforcing the negative stereotype of Indians being ruthless savages standing in the way of the expansion of the American West.

In my adult years, having taken many trips out to the American West and having visited many Indian reservations in Arizona, Wyoming, and Montana, I have come to learn of the other side of the story. The other side of history reveals that it was the U.S. Cavalry and the American pioneers who treated the Indians inhumanely and ultimately confined the Indians to reservations on unproductive soils not wanted by pioneers and settlers.

How many movies would have been financially successful had the presentation of Indians been different? Would American moviegoers have bought many tickets to cowboy and Indian movies if the Indians were portrayed as the victims of inhumane treatment by American pioneers and the U.S. Cavalry? Probably not.

History should not be viewed as accurate and factual accounts of past events. It depends on who writes the history books, what side they are on, and what their personal beliefs are and how they see the world.

During my early adult years, at a time when I was working for major American corporations, I completely embraced the corporate culture, along with having strong conservative and Republican political views. I voted the straight Republican ticket, especially during the Goldwater, Nixon, and Reagan presidential election years.

I was typical of most voters, having developed a political ideology and only caring to listen to sound bites that confirmed and reinforced my conservative, Republican beliefs.

Today I am not a Republican or a Democrat, or a conservative or a liberal. While it is not always possible, I try to make a conscious attempt to avoid having a political ideology or special interest drive my voting decisions. Each election year I assess the most important issues facing America and vote for the candidate that I think has the best grasp of the issues and the leadership qualities to effect change.

The point is that I have come to learn that a person's beliefs and actions are formed by their environment. People develop a mind-set and ideology and only see what they want to see; they only seek out and embrace information that supports and reinforces their beliefs.

In later chapters I cover how beliefs are formed, how they are used to promote special interests, and why special-interest groups and well-funded lobbyists are a threat to democracy.

INTRODUCTION

"The trouble with the world is that the stupid are cocksure and the intelligent are full of doubt." —Bertrand Russell

The twentieth century can be described as the era of the industrial revolution, with a news media offering balanced news reporting prepared by professionally trained journalists. Educating students by teaching them to memorize content and teaching to test scores was adequate for this time in history.

The twenty-first century is shaping up as the era of globalization, with a leveling of the international playing field for jobs, economic balance, and the virtual real-time global transfer of information over the Internet by social networks, Web sites, and blogs. It is also shaping up as an era where the media cable TV networks, and some radio networks in particular, are in a state of decline and risk becoming a threat to American democracy with a proliferation of strongly biased, polarizing political programs that use scare tactics and emotional generalities to stir audiences and ratings in the interest of shareholder profits.

In the last twenty-five years, the broadcast TV network affiliates, in an effort to increase viewers and ratings and offset the competition from the emerging cable networks employing emotionally charged programming and fast-paced politically biased sound bites under the guise of news, have changed the format and content of local news programs, replacing it with content heavy on crime, disaster, scandal, celebrities, and sports.

As a result, America is facing threats to its democracy and is at risk as a world leader in creating peace, prosperity, job opportunity, and a high standard of living.

The threat to our freedom in America will no longer come from countries with authoritarian dictatorships and strong militaries. Instead, the twenty-first century will see the major threats to our freedom and prosperity coming from self-interest militant groups such as terrorists, al-Qaeda, and the Taliban; from domestic self-interest non-militant entities such as well-funded lobbyists and the cable TV broadcast networks; and from international countries such as China and India gaining prominence from the advances of globalization.

Globalization will have an adverse effect on our economy and prosperity. The playing field will be leveled. Developing countries such as China, India, and the Southeast Asian countries will advance their economies at the expense of the United States, Europe, and Japan.

Technology advancements and lower overseas wages have already caused the migration of jobs and industries in America to overseas-based businesses.

In the years ahead, America will see unprecedented economic and information dissemination challenges. With exponential technological advances driving the Internet, social networks, and information dissemination, and the expanding economies of China and India brought on by globalization and the financial center of power shifting to China, America's job opportunities, prosperity, and standard of living will no longer be the model for the world to follow. America will also be faced with the crisis of becoming a strongly polarized nation as it struggles to maintain an informed citizenry that can respond to the threats posed by self-interest militant groups and self-interest non-militant entities.

Despite my tendency to lean on the side of optimism, I fear that with all this happening, America may not continue to see the prosperity and strong middle class we had during the twentieth century.

America cannot stand still and continue to cling to outdated twentieth-century beliefs, foreign affairs policies, and domestic economic policies, and still expect to compete in a twenty-first-century new world order.

With globalization and exponential technology change taking place in the twenty-first century, now more than ever we need to have an informed citizenry with critical thinking skills to ensure that America can adapt and stay out in front of the global trends that are emerging.

The biggest challenge to American democracy and freedom is now coming from internal forces, such as the rapid growth of media disinformation and the inability of our citizenry to deal with sorting out fact from fiction, generated by the rapid growth of political entertainment as opposed to credible news reporting.

The overnight emergence of the cable TV broadcast and radio talk-show programs has spawned a whole new way of thrusting special-interest disinformation on the American public. Technology advances and the need for the broadcast industry to generate profits to offset declining advertising revenue from regular scheduled family programming have changed the way Americans are fed and digest information.

An example of this is the cable TV networks and radio talk-show programs adopting a political bias that feeds and reinforces viewers with similar biases, such as MSNBC having news programs with a predominant "liberal" slant and Fox News having news programs with a predominant "conservative" slant. Not presenting balanced information on both sides of an issue is intentionally misleading and is a form of special-interest disinformation.

The huge sum of lobbyist and private campaign financing income the broadcast industry receives for political commercials and politically motivated talk-show hosts drives their programming and profits at the expense of the American public, whose listeners and viewers, deficient in critical thinking skills, ultimately succumb to hearing what they want to hear and will tune in to the programs that support their views. The deficiency of critical thinking in our country lies with our educational system, which emphasizes memorization and teaching to pass a test. Our children are being taught *what* to think, not *how* to think.

America, while having the greatest material assets in the world, has risen to become a global leader not because of its material wealth, but rather because of its greatest asset, the people of the United States.

In the twentieth century, because of the freedom of the press, highly skilled journalists and news broadcasters were able to reach every American citizen with informed, factual, non-biased presentations of the day's news and events. The American public was left to decide for themselves how to vote and support their country's best interests.

Today we are not the America of the twentieth century. America is now in danger of becoming a nation of non-thinking zombies, addicted to the political entertainers and radio talk show hosts who seek high ratings by posing as journalists and political news analysts.

A college degree or professional occupation does not necessarily guarantee or require acquiring critical thinking skills.

There is the misconception that college graduates are well-educated. Most college curriculums are content- and training-based, preparing the graduate for employment. While a number of college courses have elements of critical thinking skills, such as writing a thesis, qualitative and quantitative decision-making, and regression analysis courses that analyze data to include all variables that can affect the conclusion of the study, most college graduates have not taken any of these courses. The skills for professional careers, despite having been acquired at the undergraduate or graduate level, generally do not include courses on critical thinking or decision-making skills.

Some of the most outspoken and articulate propagators of misinformation are college graduates who justify their biased beliefs on the basis of their specialized career training and subsequent success; they now feel qualified as experts on many subjects, especially on political issues. Many of these self-proclaimed experts can be spotted spewing out quick sound bites and quoting invalid studies, surveys, and statistics with skewed data that are funded by special interest groups with a position to promote.

People who are addicted to the polarized political talk-show hosts can be identified by the language they use to make their point. Their vocabulary is filled with labels and emotionally charged words, all designed to overwhelm any serious two-way conversation. They include words and labels like patriotic, coward, liberals, conservatives, right wing, left wing, etc.

Archie Bunker was the star character of a very successful family sitcom in the 1970s and became popular for using language that was not previously used in a prime-time family show. His use of politically charged emotional labels to put down an opposing idea was viewed as laughable and entertaining. Though unknowingly at the time, this type of emotional, humorous putdown may have been the start of a whole new breed of political entertainers and a method of gaining viewers to tune in to hear political celebrities pass themselves off as credible.

One of Archie Bunker's most popular political labels was "pinko commie," which he often used when he did not have a credible thought and was backed into a corner and used that label to put down his son-in-law.

People tend to tune into programs that support their beliefs and have become addicted to cable TV talk-show political entertainers who offer simplistic sound bites and scare tactics to support their political agenda. Examples of sound bites used by political entertainers to promote a political agenda are as follows:

- Bill Maher: *"Seriously, Mr. President, this job can't be fun for you anymore. There's no more money to spend; you used up all of that. You can't start another war because you also used up the army. And now, darn the luck, the rest of your term has become the Bush family nightmare: helping poor people."*

- Ann Coulter, a frequent guest on political talk shows: *"If we took away women's right to vote, we'd never have to worry about another Democrat president. It's kind of a pipe dream, it's a personal fantasy of mine, but I don't think it's going to happen. And it is a good way of making the point that women are voting so stupidly, at least single women."*

  Our Founding Fathers established our form of democracy on the trust and belief that a well-informed citizenry would be maintained with an open market of ideas, freedom of assembly, and freedom of the press. Today newspapers and network broadcast TV are in rapid decline, and their journalistic standards are declining right along with them.

In the broadcast radio/TV industry, the balanced news, fact-oriented broadcasters like Edward R. Murrow and Walter Cronkite have now been replaced by radio/cable TV political talk-show hosts who entertain and rouse the emotions of their audiences with strongly biased, unsubstantiated opinions that frighten uninformed people into becoming loyal followers.

The major broadcast networks and tabloids exist and prosper from advertising revenue, with radio and TV talk-show political entertainers being a major source of revenue.

Add to this barrage emotionally charged disinformation from broadcast radio/TV political commercials that are paid for by lobbyists and private campaign financing, and we now have a nation of citizens who have become addicted to the quick, emotionally charged, political sound bites that offer both entertainment and political disinformation at the same time.

Sadly for American democracy, many Americans (certainly enough to sway the outcome of an election) form their political views from these strongly biased, emotionally charged sources of information.

This is all legitimate and falls within our constitutional rights of freedom of the press, but the American citizenry is ill-equipped to handle the assault of disinformation and rhetoric coming from political entertainers.

The beliefs of our citizenry become established early in the child development years and are formed from our environment and the elders who want to ensure that the next generation of adults continues to share the same beliefs.

These beliefs become deep rooted and more strongly biased unless challenged by education, and more importantly, by an education system that builds its foundation on the development of critical thinking skills.

Information and Technology Trends

"The future is embedded in the present." —John Naisbitt

Corporations, government agencies, the military, other institutions, and special-interest groups, to remain or arrive in the forefront, all seek to understand the major trends taking place and, with strategic plans in place, acquire and allocate the resources necessary to achieve their future goals.

To better understand the impact of globalization on America, we need to dig deeper into what is happening in the developing countries, what the major technology corporations are doing, and what our young generation of technology users are doing.

An excellent and widely accepted source of recent research and documentation of the exponential progression of globalization and information technology is the 2008 and 2009 versions of the video *Did You Know?*, created by Karl Fisch and modified by Dr. Scott McLeod and Jeff Brenman. This video has been viewed by many well-recognized forces of change, including military leaders in the Department of Defense (DOD), incoming congressmen, the *Economist Magazine*, countless major corporation board members and strategic planners, university presidents, and teachers.

In their videos, which are free and can be downloaded over the Internet, they report that

- If you're one in a million in China, there are 1,300 people just like you.

- China will soon become the Number One English-speaking country in the world.

- The top ten in-demand jobs in 2010 did not exist in 2004.

- We are currently preparing students for jobs that don't yet exist using technologies that haven't been invented in order to solve problems we don't even know are problems yet.

- The U.S. Department of Labor estimates that today's (in 2008) learner will have ten to fourteen jobs by the age of thirty-eight; one in four workers have been with their current employer for less than a year; and one in two employees has been there less than five years.

- One out of eight couples married in the United States last year met online.

- The first commercial text message was sent in December 1992. In 2008 the number of text messages sent and received every day exceeded the total population of the planet.

- The number of years it took to reach a market audience of 50 million was as follows: Radio thirty-eight years, TV thirteen years, Internet four years, iPod three years, Facebook two years.

- The number of Internet devices in 1984 was 1000. The number of Internet devices in 1992 was 1,000,000. The number of Internet devices in 2008 is 1,000,000,000.

- The amount of new technical information is doubling every two years. For students starting a four-year technical degree, this means that half of what they learn in the first year of study will be outdated by their third year of study.

- Predictions are that by 2049, a $1000 computer will exceed the computational capabilities of the entire human species.

- Well over 1,000,000 books are published worldwide every year. A Google Book Search scanner can digitize 1,000 pages every hour.

- Americans have access to 1,000,000,000,000 Web pages; 65,000 iPhone apps pages; 10,500 radio stations; 5,500 magazines; and 200+ cable TV networks.

- There are 240 million TVs in the United States, 2 million of which are in bathrooms. When was the last time you read a newspaper in the bathroom?

- Newspaper circulation is down 7 million over the last twenty-five years. But in the last five years, unique readers of online newspapers are up 30 million.

- 95 percent of all songs downloaded last year weren't paid for.

- Wikipedia launched in 2001. It now features over 13 million articles in more than 200 languages.

- The average American teen sends 2,272 text messages each month.

- Today 93 percent of U.S. adults own a cell phone. But one-third don't yet feel safe using it for purchases, unless we're talking about pizza.

- In February 2008, John McCain raised $11 million for his U.S. presidential bid. That same month, Barack Obama attended no campaign fundraisers. Instead, Obama leveraged online social networks to raise $55 million in those twenty-nine days.

- Among larger U.S. companies in 2009, 17 percent disciplined an employee for violating blog or message board policies.

- Twitter played an unprecedented role in sharing information during the 2009 Iranian presidential election. All mentions of the disputed election were bumped from Twitter's trending topics list when news of Michael Jackson's death broke.

- 90 percent of the 200 billion e-mails sent every day are spam.

- The mobile device will be the world's primary connection tool to the Internet in 2020.

- The computer in your cell phone today (2009) is a million times cheaper and a thousand times more powerful and about a hundred thousand times smaller than the one computer at MIT in 1965. So, what used to fit in a building now fits in your pocket; what fits in your pocket now will fit inside a blood cell in twenty-five years.

So What Does This All Mean?

"The kindergartners that start in the fall of 2007 will graduate in the spring of 2020. As architects of schools, you need to have a 2020 Vision. Your clients' and children's futures depend upon it." —Karl Fisch, Author of *Did You Know*

The media landscape continues to change at an exponential rate as a result of convergence of devices, communication, and technology. Technology has changed everything, and the old rules no longer apply.

With a changing media, people's attention span has changed, and how they communicate has changed—not in a small incremental way, but massively and disruptively.

On a moment to moment basis, we are being subjected to a massive amount of information from Facebook, LinkedIn, blogs, tweets, texting, e-mails, and Web sites and have to quickly scan to select the data we want to see. Though this tends to be disruptive, we feel compelled to tune in for fear of missing out on something important. We are now a world of people with Attention Deficit Disorder (ADD), with a compelling need to jump from one piece of information or communications media to another.

It demonstrates that the twenty-first century will continue to see exponential change in information technology and the Internet and the demise of the printed media (newspapers and magazines).

Countries, corporations, and institutions with well-informed and critical-thinking citizens, employees, and members will emerge as the new world leaders and create a new world order.

It also means that the Internet will provide a leveling of the playing field for real-time information dissemination and distribution.

This is good news for citizens under authoritarian dictatorships because they will no longer be dependent on state-controlled radio and television broadcasts for information.

However, this also can mean bad news for nations with thriving democracies, as their citizens become more exposed to strongly biased, polarized information.

BELIEFS: HOW THEY CONTROL YOUR ACTIONS

"When everyone is thinking the same, no one is thinking." —John Wooden, UCLA basketball coach

To fully grasp how we form opinions, ideologies, and mind-sets—in other words, the way we think—we need to begin with an understanding of how we develop our beliefs and how this differs from the way we develop knowledge.

Our beliefs are what we perceive and accept as true. Beliefs are not based on fact, reality, or truth. They are the way we view, understand, and interpret issues and the world we live in.

Our beliefs are influenced and formed at an early age by our environment, our role models, and everyone around us. They come from information and influences of the beliefs that are passed down by our family, teachers, religious institutions, friends, peers, books, movies, TV, work environments, social networks, political parties, government agencies, and media influences such as entertainers, radio/TV talk-show hosts, and commercials.

As adults, and especially as older adults, many people find it hard to expand their consciousness and thinking and tend to keep the same limiting beliefs they picked up during their childhood.

An example of why beliefs limit growth and expansion of thinking is a person who is a firm believer and a "know-it-all." Since they feel they know it all, they think they have nothing else to learn. Anything new presents a threat to them and makes them feel inadequate, giving them reason to deny it.

Our beliefs determine the way we live our lives and the way in which we make decisions. Beliefs drive our actions.

In order to change what's happening in your life, you have to examine your beliefs to discover why they are limiting you in getting what you want. You then need to replace them with new beliefs that will get the results you want. You won't be motivated to do something unless you believe it. If, for example, you believe you aren't good enough to be a good friend, you will not allow yourself to be a good friend.

Many people with rigid or ingrained beliefs are sure that their belief system is the right and only way and that everyone else must conform to this view.

We are not conscious of how factual our beliefs are, though they appear as representing reality to us. And because we perceive our beliefs as being real, we cannot grasp how they affect our behavior and the behavior of others.

Culture is one of the things that affects our beliefs and behavior and the beliefs and behavior of others. Culture is learned from an accumulated experience that is socially transmitted. A culture is a way of life for a group of people that drives their beliefs and behaviors that are accepted without question and are passed along by communication and imitation from one generation to the next.

Cultures are good and necessary in many ways, though they strongly influence or determine our beliefs. But to the extent that they lock us in to one way of looking at the world, we need to transcend them. We need to think beyond them.

BELIEFS AND KNOWLEDGE

According to Dr. Jay Polmar, a teacher and instructor at colleges and universities in the Southwest and Hawaii and author of *Think Right,*

> *Beliefs determine how we see, interact with, and experience the world around us. Beliefs are ideas that are formed after repetition and contemplation that are accepted as truth and reality and therefore impact how we see life.*
>
> *Belief and knowledge are often in conflict…. Knowledge is something that you and others, who might be considered experts in that field, consider to be true and there is reasonable, plausible, and provable scientific explanation for that knowledge.*
>
> *A belief on the other hand is established on personal experience or faith.*

Note that a belief is established on *personal* experience or faith. A belief is not knowledge and therefore usually not provable or disputed. The acquisition of knowledge requires an inquisitive and vigilant approach to a process that involves complex cognitive processes, perception, learning, communication, association, and reasoning.

Fortunately or unfortunately, depending on how the belief was acquired, beliefs can and do change over time. We have the freedom to change our beliefs simply by giving ourselves permission do so.

Political Beliefs

"People don't turn into conservatives because they watch Lou Dobbs or listen to Rush Limbaugh; they watch Dobbs and listen to Limbaugh because they are conservative."—Pamela Rutledge, PhD, MBA, Director of the Media Psychology Research Center

Most people develop their political beliefs during their childhood or early adulthood years and continue with little change through adulthood. The most influential sources of the formation of political beliefs are the family, schools, churches, and the media. Of these, adult family members generally wield the greatest and longest-lasting influence on a person's political beliefs.

However, as a person advances into adulthood, the influences of the family begin to wane, especially if a person becomes deeply immersed in a work environment, religious institution, or special-interest or social group with entrenched or rigid beliefs that over time replace the childhood or early adult family influences.

Angus Campbell and associates, authors of *Classics in Voting Behavior,* say

> *There is a "high degree of correspondence" between the political party an individual prefers and the party that his or her parents preferred, especially if both parents preferred the same party.*

Beliefs Become Ingrained in Your Mind

"To accept opinions is to gain the good solid feeling of being correct without having to think." —C. Wright Mills

A person can be hypnotized into believing that an apple is a grape or that a safety pin is so heavy they cannot lift it. These ingrained beliefs alter a person's perception of truth, and by doing so, alter their beliefs and actions.

In a similar way, people become hypnotized throughout their lives by the thoughts, views, and beliefs they accept from others.

When a person believes something to be true, regardless of whether or not it is true, they will act as if it is true and look for evidence to support that belief.

Beliefs can block the acquisition of knowledge. Unless some extraordinary event jolts the believers into thinking otherwise, no one will be able to persuade them that what they believe is not an accurate representation of reality.

Since many of our judgments and actions come as a result of the beliefs that have become ingrained in our minds, it is extremely important that we understand that what we consider to be true and factual may not be so at all.

Instead, we need to hold to the possibility that we have been hypnotized by false beliefs and that these beliefs may be preventing us from seeking knowledge.

The beliefs we have programmed into our minds can control not only our thoughts, but also our actions.

People have a natural tendency to accept things that agree with their current beliefs and reject things that disagree or conflict with their beliefs.

Dr. Robert Anthony, author of *The Ultimate Secrets of Total Self-Confidence* says, "The degree to which you will awaken will be in direct proportion to the amount of truth you can accept about yourself."

In other words, the degree to which you "dehypnotize" yourself from your false beliefs will be in direct proportion to the amount of truth you are willing to accept about yourself, regardless of whether you like that truth or not.

A person's belief system does not necessarily impact all of their acquisition of knowledge and their actions in the same way. For example, a person may have achieved great success in their profession, trade, or work environment by acquiring knowledge and expertise in the pursuit of excellence with an open mind, while in their personal life they may have held onto strong beliefs with a closed mind and have blocked their pursuit of knowledge.

False beliefs are not easily identified and not easily changed. This is especially true of ideological or political beliefs. Without a shift in perspective, it is very difficult to change a belief. Unless you willingly seek out and discover new evidence that contradicts your beliefs and you are willing to consider a new perspective, your belief system is unlikely to change. If, for example, you believe the earth is flat, all opposing evidence is rejected until you shift your perspective.

Do You See Only What You Want To See?

"We go around actively searching for things to see and … see mainly those things that were expected." —English neuro-anatomist Dr. J. Z. Young

A study of what optical illusions are and how we make sense of everything helps us to understand how human visual perception works.

An optical illusion is an image that differs from objective reality. Illusions trick us into perceiving something differently than it actually exists, so what we see does not correspond to physical reality.

Your brain tries so hard to make sense of everything that it often finds meaning even where there is none, thus creating optical illusions.

Psychological or optical illusions are used by scientists to demonstrate that one person can see something entirely different in the same picture than someone else. Different people looking at the image or picture will oftentimes see different things—and both will be right according to their point of view. It is possible to have different points of view based on various interpretations of things.

According to Eric Kandel, psychiatrist, neuroscientist, and professor of biochemistry and biophysics at the Columbia University College of Physicians and Surgeons, *"The organizational mechanisms of vision are best demonstrated by (optical) illusions. Illusions illustrate that perception is a creative construction that the brain makes in interpreting visual data....* *Learning does not prevent us from being taken in by these illusions."*

One of the most widely seen illusions on the Internet is the Old Woman–Young Lady illusion, which plays on what is called perceptual ambiguity. Looking at certain regions of the image, your mind will favor either the image of the old woman or that of the young lady.

The message here is that there are two sides to every story. One individual being right doesn't mean the other person has to be wrong.

This can be especially true with ideas, and we can be deceived by false beliefs and a sense of reality that controls what we want to see and interpret. Can we therefore ever be arrogantly sure of being right about ideas again?

If a person is filled with false beliefs that they regard as true, they will reject or resist any true information that conflicts with their beliefs.

According to Sam Wang, an associate professor of molecular biology and neuroscience at Princeton, and Sandra Aamodt, a former editor in chief of *Nature Neuroscience*, *"A phenomenon, known as source amnesia, can also lead people to forget whether a statement is true. Even when a lie is presented with a disclaimer, people often later remember it as true. Campaign strategists exploit it to spread misinformation. They know that if their message is initially memorable, its impression will persist long after it is debunked. In repeating a falsehood, someone may back it up with an opening line like 'I think I read somewhere' or even with a reference to a specific source."*

Human behavior is conditioned throughout life to respond in predictable ways. These responses can be mental, physical, and emotional and can cause people to automatically reject new information.

For example, a person who gets angry or defensive when you try to tell them how they could do something better or improve themselves is unlikely to learn from any beneficial information they hear because they have been preprogrammed to automatically reject it.

This creates a self-reinforcing cycle where one false belief leads to another until your mind becomes completely distorted from the true reality about yourself and the world you live in.

Rumors: The Propagation of Myths and Disinformation

"Truth doesn't always win in the marketplace of ideas. Lies spread too."
—Chip Heath, author of *Made to Stick: Why Some Ideas Survive and Others Die*

Rumors have always been around throughout history; however, with the popularity of the Internet and cable TV, a new breed of political entertainers and talk-show hosts have now become major sources of rumors and have found acceptance with people who find that they conform to their beliefs.

Some people and some groups are more inclined to accept rumors because those rumors fit with their self-interest and beliefs. This is especially the case if most of the people that conform to our beliefs or belong to our self-interest group tend to believe them too.

As Cass R. Sunstein writes in his book *On Rumors*, *"Most rumors involve topics on which people lack direct knowledge, and so most of us defer to the crowd. As more people defer, thus making the crowd grow, there is real risk that large groups of people will believe rumors even though they are entirely false."*

Rumors can also become more unshakable when deliberated among like-minded people and groups having the same beliefs.

In an experiment conducted in 2005, University of Chicago professors David Schkade, Reid Hastie, and Cass Sunstein brought together sixty citizens from two cities in Colorado to explore three of the most controversial issues of the day: affirmative action, an international treaty to control global warming, and civil unions for same-sex couples.

The people sorted into groups of like-minded people, where the people in Boulder (a predominantly liberal area) deliberated with others from Boulder and people from Colorado Springs (generally Bush country) deliberated with people from Colorado Springs. The people expressed their views in three ways: anonymously, before deliberation began; in small groups, which deliberated and tried to reach verdicts; and anonymously, after deliberation concluded. The key question was "What would be the effect of deliberation on people's views?"

The results of this experiment showed how deliberation among like-minded people can increase extremism, intensify polarization, and also squelch internal disagreement.

The Colorado experiment produced three findings:

1. That liberals, from Boulder, became distinctly more liberal on all three issues. Conservatives, from Colorado Springs, became distinctly more conservative on all three issues. The result of deliberation was to produce

extremism—and groups were more extreme than their individual members.

2. As a result of deliberation, the division between liberals and conservatives became much more pronounced.

3. Deliberation significantly decreased diversity among liberals and among conservatives.

The Colorado experiment is a case study in group polarization, showing that when like-minded people deliberate, they tend to adopt a more extreme position.

The key point here is that when presented with a rumor, internal deliberation among like-minded people or people with similar beliefs further entrenches the rumor. People become more confident and extreme in their views when they are presented with rumors that are corroborated by people with similar beliefs.

According to Marc Sageman, an independent researcher on terrorism, former member of the CIA, and consultant to the U.S. government on terrorism, *"The interactivity among a group of guys acted as an echo chamber which progressively radicalized them to the point where they were ready to collectively join a terrorist organization."*

With the power of the Internet and listservs, blogs, and discussion forums, groups of like-minded people can engage in sharing their like-minded beliefs to become more confident and extreme, ultimately leading to radicalization.

As we have demonstrated, people tend to process information and make judgments based on their beliefs and emotions, and these beliefs will become entrenched if deliberated among like-minded people. If you want people to modify their views, it is best to offer them a different view from someone they perceive as being similar in beliefs.

For example, if you are a conservative and a rumor surfaces from a liberal discrediting the conservative movement, you are quick to dismiss it. However, if a rumor comes from a conservative that discredits the conservative movement, then you are likely to give it some credence, whether it is fact or fiction.

Mind Control: Persuasion and Propaganda

"Propaganda is to democracy what violence is to totalitarianism." —Noam Chomsky

Most people see "mind control" as a bad or a dangerous thing done by bad people. This is not always the case. Often the person, institution, or group using subtle mind-control tactics is motivated by the most altruistic ideals. Mind-control techniques are widely used by advertisers, politicians, political talk-show hosts, corporations, and government agencies.

Mind-control tactics are designed to do one of two things: influence others to produce a desired response or influence yourself to make positive changes in your life.

Many mind-control tactics occur in the form of subtle messages that are relentlessly repeated and reinforced. They are on television in the form of ads or political talk-show hosts, on the Internet, in magazines and newspapers, and on highway billboards.

Propaganda is specific and is aimed at supporting a belief or an agenda. Though the message will likely convey true information, it is often partisan and paints an incomplete picture, leaving out key information that does not support the belief or agenda.

As individuals we use mind-control tactics on a daily basis to persuade others. Our thoughts and messages are used in raising our children and persuading our work subordinates and peers, friends, relatives, and others. They become embedded in our consciousness, and if allowed to go unimpeded and unquestioned, these thoughts and actions become part of our being as well as the being of others.

Mind-control tactics have the common goal of producing attitudinal and behavioral changes without being clearly visible or obvious to those undergoing the process.

Mind control is getting people to do what you want them to do by controlling or influencing the thoughts, emotions, and actions of others.

Mind control is disguised by many euphemistic names, such as persuasion, sales skills, value statements, politics, advertising, and so on. The one thing that they all have in common is the desire to change people's minds and behaviors.

The horrid truth is that, despite some people having a deep-rooted foundation of critical-thinking skills, most people tend to use subtle mind-control tactics every time they open their mouths to talk.

Corporations, government agencies, and other organizations use slogans or mottos that are rich with emotionally charged words that are designed to influence the thoughts, emotions, and actions of their employees, membership, or customers.

Three examples of the use of mottos to influence the thoughts, emotions, and actions of others are the British Special Air Service's *"Who Dares Wins,"* Morehouse College's *"To Uplift the Human Race Through Responsible Citizenship,"* and Iraq's *"Allah Is Great."*

Most people are not aware that subtle mind-control and propaganda tactics are being used on them, but in every society they are pervasive and being used all the time.

While many of these institutions would not flourish without a statement of core values and thinking habits that are based on subtle mind-control tactics, it's important to be aware of these tactics and to make certain that they do not pervade or carry over into the thoughts and actions of our everyday personal lives.

Mind-control tactics are often used over TV, the Internet, radio, newspapers, magazines, and face-to-face meetings by political talk-show hosts, politicians, special-interest groups, advertisers, and salesmen.

According to a variety of sources, including Trinity University Assistant Professor Aaron Delwiche and Investigative Journalist Marjorie Tietjen, commonly used mind-control tactics include the following:

- Subliminal Messages: Such messages are used in all forms of media—advertising, art, and so on.

- Labeling or Name-Calling: This tactic consists of attaching a negative label to a person or a thing. People engage in this type of behavior when they are trying to avoid supporting their own opinion with facts. Rather than explain what they believe in, they prefer to tear down others who hold opposing views.

- False Analogy: In this tactic, two things that are not similar are depicted as being similar. When examining the comparison, you must ask yourself how similar the items are. In most false analogies, there is simply not enough evidence available to support the comparison. An example of a false analogy is *"People are like dogs. They respond best to clear discipline."*

- Stereotyping: This tactic is used to discredit a person or cause by taking a person's single issue and placing that person into a category and attaching a negative connotation to discredit the person or issue. Common stereotype categories that promoters of misinformation attach negative connotations to include liberals, conservatives, capitalists, socialists, blacks, whites, Hispanics, feminists, hawks, doves, and atheists.

- Emotional Generalities: This tactic uses important-sounding, graphic character descriptive words that are used to create feelings of attachment or guilt. These words are often used in connection with meaningless sound bites to stir emotions. Words like "good," "evil," "patriotic," "unpatriotic," "courage," "coward," "hero," "loyalty," "justice," "integrity," and "liberty" are examples of words used to sir emotions and motivate people to join the bandwagon.

- Intimidation: This tactic is used to suggest or imply that failure to adopt the approved attitude, belief, or consequent behavior will lead to dire consequences.

- Transfer: In this tactic, an attempt is made to transfer the prestige of a positive symbol to a person or an idea. For example, using the American flag in connection with a political event or issue conveys the message that the event is patriotic and in the best interest of America.

- Using Plain Folks: This tactic uses an ordinary citizen doing ordinary activities to convince us that it is representative of America and is used to support someone or something.

- Testimonials: This tactic is easy to understand. It is when celebrities or personalities are used to support an issue or position. Whenever you see someone famous endorsing a position, you need to challenge the testimonial and ask yourself how much that person has firsthand knowledge about an issue and what he or she stands to gain by supporting it.

- Out-of-Context Data: Taking data or information out of context is used to slant a message. Key words or unfavorable statistics may be omitted, leading to a series of half-truths.

- Black or White Positioning: This tactic states that there are two choices to an issue. You are either for something or against it; there is no middle ground or shades of gray.

It is used to polarize issues and negates all attempts to find a common ground.

- Faulty Cause and Effect: This tactic suggests that because A follows B, B must cause A. Just because two events or two sets of data are related does not necessarily mean that one caused the other to happen. It is important to evaluate data carefully before jumping to a wrong conclusion. An example of a faulty cause and effect is the statement "A black cat crossed Judy's path yesterday and, sure enough, she was involved in an automobile accident later that same afternoon."

How Mind Control, Propaganda, and Persuasion Tactics Are Used

After carefully digesting these subtle mind-control tactics to promote false beliefs and misinformation and to influence the thoughts, emotions, and actions of others, it's no wonder that political entertainers and political talk-show hosts use them. They are designed to create controversy, charge up emotions, and feed and reinforce the beliefs of their viewers, who become addicted and set up their radio or TV tuner with a favorite button.

Political entertainers and talk-show hosts that are major offenders of using these tactics to promote their agenda include Rush Limbaugh, Keith Olbermann, Glenn Beck, Ann Coulter, Bill Mahr, Sean Hannity, Joe Scarborough, Jon Stewart, Marc Levin, Bill O'Reilly, and Lou Dobbs.

Major corporations, government agencies, institutions, and special-interest groups would not enjoy success unless every employee or member of that group understood and practiced the top-down mission and core values. To ensure that each member of the institution is highly motivated, enjoys high morale, and is aligned with the mission and core values, extensive training and the use of subtle mind-control tactics are needed to reinforce these beliefs.

Here are some examples of a few well-known American institutions with missions or core values that are designed to influence the thoughts, emotions, and actions of others:

United States Marines—Core Values

- Honor

- Courage

- Commitment

United Health Care—Core Values

- Integrity

- Quality

- Innovation

- Diversity

- Social Responsibility

- Generate shareholder value

- Cultivate an engaging workplace

- Work collaboratively

AFL-CIO Union—Mission

- To improve the lives of working families

- To bring economic justice to the workplace and social justice to our nation.

Ku Klux Klan (KKK)—Core Values

- For the USA

- For children

- For liberty

- For white pride

The only commonality among these organizations is that they have existed over a long period of time and have succeeded in using subtle mind-control tactics, especially in using emotional generalities to state their mission and core values.

Stating that they have used these subtle mind-control tactics for this purpose is not intended to be an indictment against the above organizations. They have enjoyed long-term prominence partly because they have been successful in creating a set of beliefs that attracts people who are like-minded. Their continued use of subtle mind-control tactics has maintained member morale, unity, and commitment to their organizations' missions and core values.

This works well for members of an organization only so long as the members of the organization remain on active status with the organization or special-interest group. However, there are risks associated with long-term employees who want to leave the organization and join another similar organization, as is the case when an employee leaves to seek a better position. Employees who have worked for too long a period for an organization with a strong "organizational culture" have developed ingrained beliefs and a "way of doing things"; they often have difficulty adapting to a new workplace environment.

For example, a creative computer software programmer who has worked more than five years for a small high-tech company in a fast-paced, loosely structured, highly creative culture, where he has grown accustomed to that culture, may have difficulty making the transition to a larger, very structured company where they have frequent employee training programs geared to company-specific operational and performance improvement issues.

People tend to interact and share common beliefs with others in the workplace or special-interest groups they are members of. This is fine, so long as we recognize the purpose of the workplace and special-interest group and guard against the organizations and people who use subtle mind-control tactics to influence our thinking and actions in our private lives. For example, we may accept a position with a new company solely for the purpose of career advancement and financial gain, and in doing so, are required to attend their training programs that influence our thoughts and actions within the workplace. We would need to put the work-related cultural issues into perspective and confine those thought-influencing actions to the workplace and not carry them over to our private lives and political decisions. Oftentimes employees with common interests tend to socialize after hours. When the only common interest is working for the same company, they are apt to extend the company's culture beyond working hours.

Failing to recognize subtle mind-control tactics and understand the purpose for which they are used can affect our ability to think clearly.

Critical Thinking—The Key

"The unexamined life is not worth living, because many unexamined lives together result in an uncritical, unjust, dangerous world." —Linda Elder

Do we want our kids to learn and grow to think and function wisely and independently, becoming their own person with choices and actions driven by facts and reason, or do we want them to become ventriloquist puppets spilling out the thoughts of others?

The best time to begin to develop critical-thinking skills is during childhood and early adult years, at the time when thinking patterns and beliefs are being thrust on them. Unfortunately most parents are not skilled in critical-thinking skills, and the elementary and high school educational systems are just now beginning to recognize the need to include these courses in the curriculum.

According to The Critical Thinking Company,

> *If we teach children everything we know, their knowledge is limited to ours. If we teach children to think, their knowledge is limitless. Our ability to succeed in life is directly proportional to our ability to solve the problems we encounter along life's journey. Tragically, elementary and secondary education is mostly*

memorization. The biggest problem facing educators today is the inability of most students to think analytically.

Good citizenship calls for the ability to think critically about the important issues facing our nation and to participate in the democratic process to ensure that our politicians are responding to an informed citizenry and acting in the best interests of our community, state, and the country.

What Is Critical Thinking?

"Critical thinking is not the same as, and should not be confused with, intelligence; it is a skill that may be improved in everyone." —D. Walsh and R. Paul

Thinking is skilled work. Critical thinking is self-guided, self-disciplined thinking that attempts to reason at the highest level of quality in a fair-minded way.

According to Greg R. Haskins's book *A Practical Guide to Critical Thinking and* Richard Paul and Linda Elder's *The Miniature Guide to Critical Thinking: Concepts and Tools,* well-cultivated critical thinkers are skilled in

- Raising vital questions and problems, formulating them clearly and precisely.

- Gathering and assessing relevant information, using abstract ideas to interpret it effectively to come to well-reasoned conclusions and solutions, testing them against relevant criteria and standards.

- Thinking open-mindedly within alternative systems of thought, recognizing and assessing, as need be, their assumptions, implications, and practical consequences.

- Communicating effectively with others in figuring out solutions to complex problems.

- Formulating and identifying assumptions that trigger or lie beneath the issue in question.

- Identifying and challenging the source of a belief, attitude, or viewpoint to ensure that the reported data or "facts" are not taken out of context and are current and from firsthand authoritative sources.

- Ensuring that emotions are kept out of evaluations of fact and logic.

- Recasting claims to make them testable.

- Recognizing when emotional language, sound bites, and labeling are being used to manipulate an argument.

- Understanding the dynamics of consensus building and group polarization and how that influences public opinion.

- Being able to apply perspective thinking; attempting to get into another person's head or walk in the other person's shoes so as to see the world the way that person sees and perceives the world.

- Evaluating the pros and cons of an issue. Who does it benefit? Who suffers from it?

- Avoiding simplistic thinking in regard to complicated issues and in coming to conclusions simply as either right or wrong.

- Being mindful of the payoffs and self-interests entailed in starting and maintaining a dispute.

While there are many elements that define what critical thinking is, there other elements that define what critical thinking is not:

- It is not thinking negatively with a predisposition to find fault or flaws. It is a neutral and unbiased process for evaluating claims or opinions, either someone else's or our own.

- It is not intended to make people think alike. Critical thinking is distinct from one's beliefs, which explains why two people who are equally adept at critical thinking but have different values or principles can reach entirely different conclusions. Additionally, there will always be differences in *perception* and *basic emotion,* which prevent us from all thinking the same way.

- It does not threaten one's individuality or personality. It may increase your objectivity, but it will not change who you are.

- It is not a belief. Critical thinking can evaluate the validity of beliefs, but it is not a belief by itself; it is a *process.*

- It does not discourage or replace feelings or emotional thinking.

According to David Ellis, author of *Becoming a Master Student, Critical thinkers: distinguish between fact and opinion; ask questions; make detailed observations; uncover assumptions and define their terms; and make assertions based on sound logic and solid evidence."*

How Do We Recognize When We Are Not Thinking Critically?

"Everyone thinks of changing the world, but no one thinks of changing himself." —Leo Tolstoy

Alfred E. Mander, in his book entitled *Clearer Thinking,* stressed the importance of conceptualizing the development of thinking as requiring training and discipline, as entailing skills that must be practiced and learned over time and through commitment. He said,

> *It is not true that we are naturally endowed with the ability to think clearly and logically—without learning how, or without practicing. It is ridiculous to suppose that any less skill is required for thinking than for carpentering, or for playing tennis, golf, or bridge, or for playing some musical instrument. People with untrained minds should no more expect to think clearly and logically than those people who have never learnt and never practiced can expect to find themselves good carpenters, golfers, bridge-players, or pianists.*

The average brain is inherently lazy and tends to take the path of least resistance. Despite having acquired critical-thinking skills, it is human nature when left unchecked to occasionally lapse and fall into the trap of propagating rumors and false beliefs or unconsciously act on ad campaigns that are intended to promote the purchase of products or services. These occasional unchecked lapses tend to become habits.

In his article "Overview of Critical Thinking," James J. Messina describes some of the habits of people who have an absence of critical thinking:

- Forwarding e-mails—There are times when it becomes very tempting to blindly pass on or reproduce information, sound bites, or jokes that fit into our beliefs. One of the largest sources of passive and reactionary information is the abundance of e-mail messages that contain jokes and other misinformation designed to be clever and informative and promote rumors and stereotyping.

- Accepting justifications at face value—Even though we are part of a corporate or government workplace; are affiliated with a social, religious, or political group; or endorse a political candidate, we need to guard against blindly accepting information from these sources.

- Unconsciously reacting to TV commercials—Do we find ourselves tempted to unconsciously react to and/or purchase the products advertised?

- Blindly accepting information, saying that "if the textbook says it, it must be so"—Many nonscientific or mathematics textbooks over time are revised as new information surfaces. History books, for example, are written from the point of view of the writer and are not necessarily accurate. Survivors of different sides of a war or conflict often portray the events of the war very differently.

Critical Thinking Is an Educational Issue

"Professors teach what to think, not how to think" —The Daily Campus

In the twentieth century, largely driven by the industrial revolution, America's elementary and high schools and to a large extent our universities were mass-educating students with content- and knowledge-based curriculums to become workers in an industrial economy. The media for the most part was focused on offering fact-based reporting and employed professionally trained journalists and reporters to deliver the news and political opinion. Educating students by teaching them to memorize content and knowledge was perfectly adequate for that time in history.

Now that the industrial revolution and the twentieth century have become history, and with globalization, the exponential rate of technology advances, and the media barrage of political disinformation that characterize the twenty-first century, we need to reinvent how we educate our kids.

Today students are for the most part technology- and knowledge-savvy, but not yet thinking-savvy. To effectively function in the twenty-first century, our educational system must produce thinking-savvy students, and that requires the use of critical thinking skills.

EDUCATION VERSUS TRAINING

"Education's purpose is to replace an empty mind with an open one."
—Malcolm S. Forbes

In the learning process, critical-thinking skills are necessary as a foundation for a highly successful education process, and to a lesser extent for the training process. Education teaches *how* to think, while training teaches *what* to think.

Just because a professional has gone to college and has become successful in their field of expertise, there is the misconception that they are well-educated. Depending on the discipline they pursue, this could be the case, but more often their expertise and success in a career or the workplace is the result of training they received at the undergraduate or graduate level.

Education is focused on the development of the mind and the intellect and on becoming a well-rounded critical thinker. The main purpose of critical thinking is the enhancement of an individual's ability to use his mind and function at the highest level of thinking. While a highly educated person is often more employable, education is not about getting a job.

Training is focused on the learning of specific skills. It is preparation for competency in an occupation, talent, or vocation. It is not education.

Training is a learning process that develops specific social or occupation skills. Trade schools, colleges (most majors), corporations, and government agencies offer training and indoctrination programs without critical-thinking courses. These programs are designed to either teach new occupation skills or develop competencies to qualify for specific jobs or careers.

The college-educated professional careers that are the exceptions (because they include critical thinking skills in their curriculum) are research, economic-based occupations, journalism, and occupations requiring extensive analysis of statistical data.

The difference between training and education can be illustrated with this parent and daughter exchange: What would be your reaction if your fifteen-year-old daughter told you that she was going to take a sex *education* course at high school? Would your reaction be the same if she told you she was going to take part in some sex *training* at high school? Not likely.

In America's school systems, the distinction between education and training can be very narrow. To become a skilled and certified accountant, for instance, requires years of training before one can be considered employable as an accountant. Training, therefore, has the end result of employment rather than the development of the mind and one's thinking ability.

The Need to Teach Children How to Think

Kathleen Hall Jamieson, Director of the Annenberg Public Policy Center of the University of Pennsylvania, says, *"As partisan outlets proliferate, students raised on faux news will enter our classrooms cocooned in their own biases and conditioned to mistake ridicule for engaged contention."*

The rapidly changing conditions of the twenty-first century and beyond require new educational outcomes. Old standards of measuring results based on standardized tests of content, though still valid, cannot be the sole means by which we judge the academic success or failure of our students.

The educational system in America has achieved excellent results in teaching students content and *what* to think and in preparing for tests. What teachers must now do is also teach students *how* to think.

Critical thinking is not only an important issue with our school system in the United States; it is also an issue with corporations and even the military. According to George A. Emilio, a major in the U.S. Air Force and author of *Promoting Critical Thinking in Professional Military Education,* *"The increasing complexity and changing nature of the human environment necessitates improved critical thinking skills. These skills have not been taught as part of traditional education nor are they typically stressed in the Air Force educational system."*

An educational curriculum with a foundation of critical-thinking skills makes certain that schools and universities are not only turning out good plumbers, good lawyers, and good engineers, but also turning out good citizens.

Unless we examine and introduce change to our educational curriculum to create a foundation of critical-thinking skills, we will be remiss in graduating well-informed citizens who have been taught how to think, and American democracy will be at risk.

According to Robert Neumann, author of *American Democracy at Risk,*

> *Democratic citizenship involves more than casting a ballot on Election Day. It involves a disposition for social responsibility and civic engagement; it involves participation in groups concerned with advancing foundational principles of liberty, justice, and equality and with improving human welfare and the environment of the country and planet. Effective citizenship requires critical habits of mind and the ability and inclination to deliberate and debate conscientiously on matters of social importance. What is needed is a more holistic approach to democratic health, and a central dynamic of that approach is education, more specifically, social education.*

Children are our future, and unless we educate and prepare them to become critical-thinking, informed citizens who are not swept up by media hysteria, scare tactics, and emotionally charged rhetoric, American democracy is at risk.

According to Steven D. Schafersman's article "An Introduction To Critical Thinking,"

> *All education consists of transmitting to students two different things: (1) the subject matter or discipline content of the course ("what to think"), and (2) the correct way to understand*

and evaluate this subject matter ("how to think"). We do an excellent job of transmitting the content of our respective academic disciplines, but we often fail to teach students how to think effectively about this subject matter, that is, how to properly understand and evaluate it. This second ability is termed critical thinking. All educational disciplines have reported the difficulty of imparting critical thinking skills.

While our educational system is best equipped to teach kids critical-thinking skills, parents can contribute by providing a nurturing environment in helping their children learn how to think.

Jessica Pegis, author and learning resource instructor for kids and educators, wrote an article "5 Ways to Develop Your Child's Critical Thinking Skills," in which she says:

You can nurture critical thinking in kids by trying out some of these ideas:

1. *Ask and observe.*

2. *Ask: How do you know that?*

3. *Talk about good and bad reasons.*

4. *Repeat: Information does not = truth.*

5. *Show respect for the other side.*

THE DECLINE OF THE BROADCAST TV NEWS MEDIA

"Pressures by media companies to generate ever-greater profits are threatening the very freedom the nation was built upon ... It's not just the journalist's job at risk here. It's American democracy. It is freedom." —former CBS News anchor Walter Cronkite

An effective democracy must have a media system that observes the highest journalistic standards and performs two critical functions:

1. The media must perform a watchdog role. They must rigorously challenge and hold the people in power and those who want to be in power accountable, both in the public and private sector.

2. The media also must present balanced news from reliable sources and effectively present it on a wide range of important current social and political issues.

In the novel *Nineteen Eighty-Four*, George Orwell said, *"Who controls the past controls the future: who controls the present controls the past."*

Up until the latter part of the twentieth century and until the availability of TV in everyone's home in America, people received their news from the daily newspapers. Journalists and other people depended on the printed medium to convey news, but poor journalism could be reflected in printed media.

Today, corporations, special-interest groups, and politicians with large public relations and marketing budgets have come to understand the power of the message and the power of the medium or media that get the message out.

Bonnie Anderson wrote in her book *News Flash: Journalism, Infotainment and the Bottom-Line Business of Broadcast News,*

> *The problem with television news isn't about the Left versus the Right—it's all about the money. From illegal hiring practices to ethnocentric coverage to political cheerleading, the American broadcast conglomerates' pursuit of the almighty dollar consistently trumps the need for fair and objective reporting. Along the way to the bottomline, the proud tradition of American television journalism has given way to an entertainment-driven industry that's losing credibility and viewers by the day.*

Journalism is the investigating and reporting of news. The basics are the 5 W's: the Who, What, Where, When, and Why of a story. This was the way news reporting in the TV broadcast industry was provided during the glory days of the fifties, sixties, and seventies, with famed newscasters such as Edward R. Murrow, Walter Cronkite—voted the "Most Trusted Man in America"—Chet Huntley, and David Brinkley.

During the last twenty-five years of the twentieth century, in an effort to ward off competition from the emerging cable networks and to increase evening news viewers and ratings, the major broadcast TV network affiliates began to change the format and content of local news programs and replace it with content heavy on crime, disaster, scandal, celebrities, and sports.

Despite his being highly influential in the TV broadcast industry, Walter Cronkite failed in his attempt to expand national news to one hour. The local network affiliates, being highly profitable with their local news satisfying viewer thirst for crime, disaster, scandal, celebrities, and sports, won out over Cronkite's efforts to expand the national news.

In 1987 the Federal Communications Commission abolished the Fairness Doctrine, which required the public broadcast license-holders to present important issues to the public and to afford reasonable opportunities for the presentation of "contrasting perspectives" on those issues.

This action was taken by the FCC despite the Supreme Court upholding the constitutionality of the Fairness Doctrine. The Supreme Court's decision only added to the controversy. The print and broadcast media were inherently different, it ruled. The court said that in the broadcast media, , *"it is the right of the viewers and listeners, not the right of the broadcasters, which is paramount ... it is the right of the public to receive suitable access to social, political, esthetic, moral, and other ideas and experiences which is crucial here."*

The Federal Communications Commission justified the decision to abolish the so-called Fairness Doctrine on the basis of government regulation of broadcast speech being in conflict with the first amendment.

But was there another motive for the FCC to abolish the Fairness Doctrine? At the time, there was pressure by one of the strongest lobbyist groups in Washington, the National Association of Broadcasters (NAB)—the association that represents the broadcast industry before the FCC—to abolish the Fairness Doctrine.

Sheila Kaplan, prize-winning investigative reporter, television producer, and lecturer in political reporting at the University of California Berkeley Graduate School of Journalism, wrote the following:

> *"The Fairness Doctrine has such overwhelming support on Capitol Hill, it's a recognition of the tremendous power of electronic media that it didn't go through," said Charles Ferris, a former FCC chairman-turned-lobbyist. Ernest Hollings, former Senator from South Carolina, agreed. "Our broadcaster friends are the most powerful I know of…," he told the Senate. "They can change votes right and left, and that is quite understandable. We live and breathe by TV, and that is our reelection. If the local broadcaster calls, you are going to do him a favor. You are not worried about a veto by the president." With the advent of cable television, and knowing they'd be heavily regulated by Congress and the FCC, cable operators burst on the scene with a sizable PAC and generous honoraria, starting a bidding war of sorts with broadcasters. Each tried to outdo the other in campaign contributions, speaking fees, and all-expense-paid convention trips for their favorite congressmen to places like Hawaii and Las Vegas, seeking, as it is euphemistically known, to "participate in the political process." Between 1985 and August 1988, the National Cable Television Association donated $446,240 to*

federal candidates. The NAB gave $307,986 during the same period. Last year (1987), the $114,300 the NAB disbursed in speaking fees was more than any trade group except the American Trucking Association.

The repeal of the Fairness Doctrine became an opportunity for corporate owners of TV news programs to trim their news budgets—and it was followed by a decline in the quality of balanced news broadcasting. CBS's Washington bureau, which employed twenty-one correspondents at its peak under Cronkite, shrank to nine by the end of the twentieth century.

In his keynote address to Columbia University students in 2007, Walter Cronkite said, *"Business people need to understand that ownership of a news company involves special, civic responsibilities. Consolidation and cost-cutting may be good for the bottom line in the short term—but it isn't necessarily good for the country or the health of the news business in the long-term."* Cronkite also expressed concern about the press maintaining its obligation to be committed to public service. He added, *"What best would serve the country and the free press is to encourage ownership by entities that are dedicated to public service—companies that invest for the long haul and will serve their communities rather than just ever-greater profits. America is a powerful and prosperous nation. We certainly should insist upon—and can afford to sustain—a media system of which we can be proud."*

The repeal of the Fairness Doctrine also opened the doors to politically biased talk radio and cable TV networks that are no longer required to offer a balanced perspective or opposing views on important public issues and especially on political issues of national importance.

With the rapid growth on cable TV networks in the twenty-first century of strongly biased political talk-show hosts, we are experiencing a decline in literacy as a result of the decline of newspapers. The cable TV news industry has succeeded in capturing the viewer's attention with video laden with fast-paced, shocking, emotionally charged images and content.

While cable TV news broadcasts have become the major offenders to pass off strongly biased political content as news, over the last fifty years, the major TV news broadcasts have steadily reduced the evening news to thirty-minute programs of disconnected sound bites.

Cable TV news has become a sideshow for loudmouths and know-nothings. It's all about the best way to get ratings—by keeping the pot stirred. What do you think will get higher ratings: a calm discussion of all the issues on foreign policy, or a highly charged political entertainer using emotional scare tactics presenting only one side of the story?

The cable TV news talk-show format that dominates cable's prime time has become ideologically polarized. Depending on their polarized beliefs, viewers tune into the network to get a dose of the latest news on issues to reinforce their beliefs.

The two major cable TV news offenders providing polarized political entertainment are Fox News, slanted to the "right," and MSNBC News, slanted to the "left." As cable TV news viewership grows, it becomes even more ideologically polarized because it has learned that its success has come by giving the viewers the biased information that they want to hear, not balanced reports.

Carl Bernstein, the *Washington Post* reporter who along with Bob Woodward broke the story of the Watergate break-in that led to the resignation of President Richard Nixon, refers to the current state of TV broadcast news as

> *Sensational journalism as public discourse turned into a kind of a news "sewer" which is perpetuating an "idiot" culture ... That type of news panders to the public and insults their intelligence, ignoring the context of real life. Our political system is in a deep crisis; we are witnessing a breakdown of the community that has in the past allowed American democracy to build and to progress. Surely the advent of the talk-show nation is a part of this breakdown. Some good journalism is still being done today, to be sure, but it is the exception and not the rule.*

Our cable TV news media conglomerates are failing to provide the unbiased, balanced information to make our democracy work. They don't challenge people to think; instead they tell people what to think. They turn debates about serious issues into shouting matches of polarized views and trivialize democracy in the interest of viewer ratings.

Unless the future voters in America become skilled in critical thinking, the health of American democracy will become contaminated with citizens having strong polarized beliefs fed by their addiction to the politically biased cable TV and radio talk-show networks.

The News Media and Journalism

"The quality of democracy and the quality of journalism are deeply entwined."
—television journalist Bill Moyers

During the fifties, sixties, and seventies, about 6:30 p.m. at least two or three nights a week, about half the country could be found watching the evening news on one of the three major networks. The news broadcasts tended to be fairly balanced, with highly professional anchormen such as Edward R. Murrow, Walter Cronkite, Chet Huntley, and David Brinkley.

The viewers had little choice; with only four or five channels to choose from, there wasn't much else on.

This all changed in the eighties and nineties with cable TV and the Internet, and the choices became overwhelming. Viewers had hundreds of channels to chose from, or they could forsake TV altogether and entertain themselves on the Internet.

As cable TV news continues to wow viewers with emotionally charged news, partisan spin doctors, and snap judgments about governance, and with cable TV companies now taking over commercial broadcast networks, such as Comcast acquiring NBC in December 2009, are we faced with the only remaining source and hope for balanced news on TV being public broadcasting?

With cable TV news rapidly growing and the media in America undergoing unprecedented corporate consolidation, sources of balanced news reporting are drying up, and at the same time so is the health of American democracy.

In the early 1980s, there were approximately fifty media conglomerates dominating all media outlets, including television, radio, newspapers, magazines, music, publishing, and film. In the year 2000, just six corporations dominated the media in America, and the trend of corporate media consolidation shows no signs of abating.

Kevin Howley, associate professor of media studies at DePauw University, Greencastle, Indiana, reported that, *"The corporate consolidation of the news industry—with the attendant demands for cost cutting on one hand and profit maximization on the other—has all but extinguished any semblance of a free press."*

Public Broadcasting

Public radio and TV broadcasting in America has never achieved the success and viewer interest that has been achieved in Japan, Canada, and Western Europe. The distinction between America and these other countries is that outside of America public broadcasting has been well-funded and designed to serve the entire population.

Public broadcasting in the United States is in deep financial trouble. Some stations, suddenly deprived of state financial support, could end up going dark. Budgets are being trimmed and jobs are being eliminated. Desperate pleas are going out to a recession-squeezed public to increase their financial contributions to the public broadcasters.

Cutbacks in funding from federal and state governments have occurred in recent years and have been fueled by ideological debates as to their justification.

In 2002 Willard D. Rowland, Jr., PhD, and President, Colorado Public Television, KBDI-TV, Channel 12, reported that

> *Federal funding for public media has always been contentious in the U.S. It lies at the heart of American ideological debates over the state of the arts, education, and communication (the "culture wars") and First Amendment issues about the role of government in such matters. As a result, even as federal funding for public broadcasting tended to increase, it had been periodically reduced and regularly subjected to serious threats of elimination altogether. Such episodes occurred in the early 1970s, the early 1980s, and again in the mid-1990s.*

The justification for public broadcasting is that it exists to fill voids and provide coverage of interests for which there are no markets. Public broadcasting supplies programming content on topics and interests that have social benefit that would otherwise not be broadcast due to unprofitability.

Public broadcasting in America is not new and emerged alongside commercial broadcasting. Most early public stations were operated by state colleges and universities, and were typically operated under the schools' cooperative extension services.

In 1969, while commercial broadcasting was still enjoying a near oligopoly market share, the Nixon administration proposed—and Congress enacted—the first federal funding bill to support development of what is now known as public broadcasting. While the financial support was meager, it established a formula to help fund public broadcasting without creating conflicts of interest or undue government control.

Under Section 396 of Title 47 of the United States Code, Congress enacted legislation in 1969 to permit the establishment of nonprofit corporations to encourage the growth and development of public radio and television broadcasting, including the use of such media for instructional, educational, and cultural purposes. It also encouraged the development of programming that involves creative risks and that addresses the needs of unserved and underserved audiences, particularly children and minorities.

In America, public radio and TV broadcasting has had to struggle with being underfunded and restricted to providing programming that is not commercially viable. Without the fast-paced, emotionally charged programming that appeals to mainstream America, and with Congress having oversight to ensure that public broadcasting does not engage in ideological dialogue beyond that provided by the commercial broadcasters, the programming content is relatively bland and has had appeal only with fringe audiences.

Public broadcasting in America is a niche service and not a comprehensive service as in some European countries. Public broadcast television in America, from the sixties on, has also had to struggle with severe criticism from conservative politicians and think tanks, which allege that its programming has a left-wing bias.

Whether this is true or not, it is unlikely that any broadcast medium will be able to be all things to all people. The important issue here is that there is a set of editorial standards in place to provide balanced reporting.

The Public Broadcasting Service (PBS), describes their editorial standards on their Web site as follows:

> *PBS's reputation for quality reflects the public's trust in the editorial integrity of PBS content and the process by which it is produced and distributed. To maintain that trust, PBS and its member stations are responsible for shielding the creative and editorial processes from political pressure or improper influence from funders or other sources. PBS also must make every effort to ensure that the content it distributes satisfies those editorial standards designed to assure integrity.*

The funding for public radio and TV in America comes largely from corporate grants and underwriting, and from private donations and subscriptions, with less than 15 percent coming from government subsidies. By international standards of public broadcasting, this is more nonprofit commercial broadcasting, subsidized by the corporate sector, than it is public broadcasting.

Despite American public broadcasting facing many restrictions and challenges, it continues to develop a following of viewers tuning in for balanced news reporting and arts programming.

There are two major public service broadcast networks in the United States: Public Broadcasting Service (PBS) and National Public Radio (NPR). PBS is a private, nonprofit corporation whose members are America's public TV stations, with 356 member stations and an audience of 59 million. NPR is an international producer and distributor of noncommercial news, talk, and entertainment programming. A privately supported, not-for-profit membership organization, NPR produces and distributes programming that reaches a combined audience of 26.4 million listeners weekly. NPR member organizations operate more than 900 stations nationwide. (This information was furnished by PBS and NPR.)

For democracy to flourish, people need unrestricted access to independent, diverse sources of balanced news and information. PBS and NPR, with editorial policies that stress balanced reporting, offer an alternative to the politically biased cable TV and radio talk-show networks.

According to Gerald Caplan, former co-chair of the Task Force on Canadian Broadcasting Policy,

It is useful to remind ourselves that free expression is threatened not just blatantly by authoritarian governments and all those in the private sector who fear public exposure, but also more subtly by the handful of global media conglomerates that have reduced meaningful diversity of expression in much of the globe.

> A necessary condition for true democracy is the widespread availability and access to objective information on all sides of a problem or issue in order to resolve the contradictions in opposing views or ideas.

As the number of viewers with critical-thinking skills hopefully grows, and with people being increasingly shut off from getting balanced news from the cable TV networks, more viewers are expected to tune in to public broadcasting, assuming of course the funding remains in place for the public broadcasting networks to continue their operations.

Internet, Journalism, and Balanced News

"Internet reporting is promiscuous, and Internet learning is individuated. The Web may be worldwide, but it encourages homophily—our tendency to seek more of what we already know and to associate with like-minded souls." —Harry R. Lewis, Professor of computer science at Harvard University.

With the exponential rate of change of technology continuing, what will happen to radio, TV, and the print media?

The new generation of news consumers, the young adults who have grown accustomed to computers and are Internet- and blog-savvy, are using the Internet as their major source of getting news and information about local and global events.

The real question is to what extent will this new generation seek out balanced news. Will they simply run keyword searches to find biased Web sites and blogs with information to reinforce their beliefs?

A real concern is that today anyone with a cell phone camera or an Internet connection, with a little Web site or blog development knowledge, can now be a "reporter." With the Internet, the public can no longer be passive consumers of news. We must increasingly become critical thinkers and our own gatekeepers of information. Similar to cable TV, the reporting we see on the Internet is not the work of journalists, but of polarized political writers with biased agendas.

Chicago law professor Cass Sunstein writes in her 2007 book *Republic. com 2.0*,

> *We are creating enclaves of like-minded people: historically, narrow-interest groups have fueled social progress, like the civil-rights movement—but also cults and Nazism. There is a general risk that those who flock together, on the Internet or elsewhere, will end up both confident and wrong.*

However, others view the Internet as a means of giving viewers access to new ideas. Al Gore wrote in his latest book, *The Assault on Reason, "The Internet is democracy's last, best hope, a way of opening the world to free-flowing ideas."*

This is certainly true in the third world and in countries where dictatorships exist. The Internet thus provides hope for free societies to emerge where freedom did not exist before the Internet.

Perhaps the Internet can be viewed in terms of globalization as a force where the peoples and countries of the world will participate in the leveling of the playing field, not just in economic matters but also in matters of freedom and democracy.

Will the leveling of the playing field of global societies mean that American democracy will suffer as a result of nonprofessional journalists dominating the information on the Internet with their strong, polarized views at the expense of balanced news?

It's easy to believe that journalism is a dying profession in America and that these are the worst of times.

Journalism has thrived with the distribution of its content through the media of newspapers, magazines, network TV, and local TV news. American democracy, with a thriving journalism profession and media distribution outlets, has also thrived throughout history.

As audiences contract in each of the distribution media, the future of journalism as we know it and the future of our society are at risk.

Throughout 250 years of American history, we have come to rely on factual and balanced reporting in newspapers to uncover problems in government, business, and society that need to be remedied. This kind of public-service journalism has kept government officials honest, contributed to the punishment of evildoers, and rallied public support for social change.

Some groupings, namely business, sports, entertainment, and technology, have already found it profitable to generate sufficient online advertising revenue to pay writers and video producers to create original content. That's because these are groupings with established bases of advertisers who want to reach the audiences interested in that specific kind of content.

But who will produce professional journalism-based content if big companies withdraw their support? And can investigative reporting continue to have an impact if the audience for traditional media continues to shrink?

Will the Internet provide the outlets and medium to support the creation of original journalism? While every traditional media company has a Web site, they serve more as complementary distribution channels for already-created content than as a medium for original journalism. For example, while the *Wall Street Journal* charges for an online subscription, the *Washington Post* and many other newspapers offer free online subscriptions where their newspaper can be delivered directly to your computer.

Will young audiences be able to maintain their attention and interest for balanced, in-depth, online news reporting when other forms of highly charged, fast-paced, polarized video and content posing as news may be more to their liking?

These are all questions that professional journalists and the media are seeking solutions for.

With Internet journalism still in its infancy, it will take time for journalistic roles and attitudes to change. But there are already signs that news organizations are reinventing their relationship with the audience and tapping into the participatory potential of the Internet to recast journalism.

An early trend is with individual journalists who no longer find a future with the large media companies, going out on their own and using the Internet successfully to market themselves with Web sites, e-mail, blogs, social media, and more. These ventures are financed through Google ads, affiliates, and donation solicitations via PayPal. This all suggests that the power in the media is shifting to the individual journalist and away by degrees from journalistic institutions.

A disturbing trend is that the news press and journalists are becoming less of a factor in political campaigns. According to a study conducted by the Pew Research Center's Project for Excellence in Journalism, "The State of the News Media—2009, *"In 1992, the* Washington Post *produced 13 major profiles examining the past record and biography of the eventual winner of the race. In 2008, the paper's ombudsman found, it produced three. At the* Los Angeles Times, *the number of such enterprise stories about the winning candidate fell by two-thirds. Many factors have contributed to this less pro-active press. Smaller newsrooms leave people less time for enterprise."*

With the rise of news and information being provided online and in real time, journalists appear to be well-placed to profit from this trend. The method of capturing a viewer's attention with text links is not much different than editing a newspaper or magazine.

Media companies are now beginning to use real-time social networks for managing staff and news content. The *New York Times* has created a Twitter list of everyone on its staff, and the *Los Angeles Times* and CNN have created lists on categorizing Twitter celebrities. The Internet blogger Huffington Post has introduced Twitter lists on its Web site to create groups for real-time updates.

These trends suggest that there is a revenue stream to support professional journalists and media companies with original, balanced content on the Internet. But can the Internet fulfill the public-service role that newspapers, network TV news, and local TV news, with their huge advertising revenue base, large audiences, and large staff of professional journalists, have enjoyed until recently?

To preserve the integrity and widespread availability of balanced news, the federal government is conducting deliberations over the future of journalism as printed newspapers, network TV, and other traditional media outlets suffer from Americans' growing reliance on cable TV and the Internet.

These deliberations are a series of workshops that began in 2009 and are scheduled to continue into 2010. They are calling upon leading executives from both traditional and new media, including Len Downie, former executive editor of the *Washington Post*; Rupert Murdoch, chairman and chief executive of News Corporation; Arianna Huffington, cofounder and editor-in-chief of the *Huffington Post* Web site; and executives from Google Inc. and Yahoo Inc.

As the chairman of the federal agency driving this initiative (the FTC), Jon Leibowitz said, *"News is a public good; we should be willing to take action if necessary to preserve the news that is vital to democracy."*

Although it is not yet known what recommendations will be made to the FTC regarding the future of journalism with printed newspapers and traditional broadcast network TV, there is a follow-up planning workshop scheduled for the spring of 2010. Among the options being considered are tax law changes that would allow media companies to earn tax credits or become tax-exempt entities and copyright law changes that would force search engines and other online aggregators to compensate media companies for the content they produce.

Social Networks and Journalism

Social networks like Twitter, Facebook, LinkedIn, and Digg produce information, not journalism. Journalism requires discipline, fact-finding, analysis, explanation, and context. While many of these social networks have become a forum for making friends and sharing bumper stickers, photos, and causes, their use by individual journalists has been primarily as a way to pick up on newsworthy topics and ideas to jump-start stories.

One of the benefits of social media is as a virtual real-time two-way communications forum where journalists and their audience can now be part of the process and have a space to voice their opinions. This is healthy in a democracy, and the idea of uncensored, free speech is key to this system. These tools help this process.

A recent example of the place of social networks in furthering the cause of free elections occurred during the alleged fraud of the Iranian 2009 presidential elections. Social networks were used by students to spread their defiance online. Labeling the spontaneous antigovernment demonstrations as a "Twitter Revolution" has already become something of a cliché.

Despite crackdowns by the Iranian government to shut down and block Facebook, Twitter, and other Web sites, Twitter users found ways to get around government snooping. Internet-savvy Twitter users shared ways of getting around the blocking with other users, such as programming their Web browsers to contact a proxy, an Internet server that relays their connection through another country.

The very nature of Facebook, Twitter, and LinkedIn as online forums for keeping in touch with friends, family, and business associates is also what makes them a very powerful tool for the oppressed. With social networks being an international forum for people of similar beliefs and interests to share, perhaps their greatest threat to balanced news reporting is the reinforcement of false beliefs and the spread of rumors to people who want to associate with people and groups with similar beliefs.

EPILOGUE

"Teachers and students must always remain free to inquire, to study and to evaluate, to gain new maturity and understanding; otherwise our civilization stagnates and dies." —Chief Justice Earl Warren, Sweezy v. New Hampshire, 1957

There is little that can be done or that we would want to do with regulation or control of the cable TV news and the Internet. The first amendment protects our freedom of speech rights:

Congress shall make no law respecting an establishment of religion, or prohibiting the free exercise thereof; or abridging the freedom of speech, or of the press; or the right of the people peaceably to assemble, and to petition the Government for a redress of grievances.

Our hope for the future and for a thriving American democracy lies with our children and with our education system. Our education system, from elementary school up to and including college, needs to reorder its priorities on what needs to be taught to our children. We need to place emphasis on the teaching of our children how to think over what to think.

Aside from food, water, and shelter, the one resource that a person will most need in life is an education. Education is the only one of the four resources that can help ensure a person's ability to provide for their future well-being. An education with a solid foundation in critical-thinking skills gives students the ability to not only understand what they have read or been taught, but more importantly, to build upon that knowledge and, by being able to take ownership of their thoughts, to identify and discard propaganda, rumors, and influence from others with an agenda to promote.

With beliefs formed at an early age, our children need to begin to develop an understanding of the principles of critical-thinking skills and take ownership of their thoughts, beginning in the fourth grade and continuing through high school and college. Critical-thinking skills can evaluate the validity of beliefs, but they are not a belief by themselves; they are a *process*.

The twenty-first century is shaping up to be an era of exponential change and advancements in information technology, bringing about unprecedented global dissemination of real-time information.

Some of this is beneficial, as it is a threat to authoritarian dictatorships and their inability to cope with the free flow of information and news.

However, this era also poses threats to American democracy and America's position as a global leader.

The print media and journalistic-driven evening news programs on the major broadcast networks such as ABC, CBS, and NBC are being replaced by cable TV news networks and Internet Web sites, blogs, and social networks. The cable TV networks in particular, with their polarized, emotionally charged political talk shows, are attracting

Americans looking for reinforcement of their political beliefs. As a result, America is becoming a nation of people dependent on being told what to think, not how to think.

Authoritarian dictatorships survive with an infrastructure of control over the school systems and TV and radio broadcast networks that teach and tell the people what to think.

For democracy to survive and flourish, our school systems and radio and TV networks must be oriented toward teaching and encouraging people *how* to think.

The threat to freedom in America will no longer come from countries with authoritarian dictatorships and strong militaries. Instead, the twenty-first century will see threats to our freedom and prosperity coming from self-interest militant groups (terrorists) and self-interest non-militant special-interest groups and political lobbyists.

Propaganda and rumors have been around throughout history. With the rapid technological advances of the Internet and the popularity of Facebook, Twitter, Web sites, blogs, and so forth, and with cable TV with its breed of political entertainers and talk-show hosts out to get ratings, rumors are more widespread than ever and have found acceptance and reinforcement with people who find that they conform to their beliefs. People are more inclined to accept rumors because those rumors fit with their self-interest and beliefs.

This generation's children are our future. They are our hope for the preservation of a healthy democracy. There is an urgent need to prepare our children for adult life with a foundation of critical-thinking skills. We must instill them with the responsibility of taking ownership of their thoughts. As parents and educators, we have an obligation to prepare our children not only for their own future well-being, but also for the future well-being of America, by giving them the gift of critical-thinking skills necessary to contribute to a healthy democracy.

"You want to prepare your child to think as he gets older. You want him to be critical in his judgments. Teaching a child, by your example, that there's never any room for negotiating or making choices in life may suggest that you expect blind obedience-but it won't help him in the long run to be discriminating in choices and thinking." —Lawrence Balter, psychologist, in *Dr. Balter's Child Sense,* 1985

Bibliography

Anderson, Bonnie. *News Flash: Journalism, Infotainment and the Bottom-Line Business of Broadcast News*. San Francisco, CA: Jossey-Bass, a John Wiley imprint. 2004.

Anthony, Robert. *The Ultimate Secrets of Total Self-Confidence*. New York, NY: The Berkley Publishing Group. 1979

Bernstein, Carl. "The Idiot Culture." *The New Republic*. June 8, 1992. <http://www.worldji.com//Bernstein.pdf>

Browne, M. Neil, and Stuart M. Keeley. *Asking The Right Questions: A Guide to Critical Thinking*. Upper Saddle River, NJ: Prentice Hall, 1998.

Campbell, Angus, Philip Converse, Warren Miller, and Donald Stokes. *Classics in Voting Behavior*. Washington, DC: Cq Press, 1992.

Caplan, Gerald. Quoted in "Media Conglomerates, Mergers, Concentration of Ownership." <http://www.globalissues.org/article/159/media-conglomerates-mergers-concentration-of-ownership>

Critical Thinking Co.™ Staff. *The Critical Thinking Co.* May 2005.
<http://www.criticalthinking.com/company/articles/
teaching-critical-thinking-skills.jsp>.

Cronkite, Walter. *Media Reform: Is It Good for Journalism?* Columbia
University Keynote Address, 2007 <http://www.journalism.
columbia.edu/cs/ContentServer/jrn/1175295260607/
page/1175295260590/simplepage.htm>

Cronkite, Walter. *A Reporter's Life.* New York: Random House
Publishing Group, 1996.

Delwiche, Aaron. "Propaganda." September 29, 2002. *The Institute
for Propaganda Analysis.* <http://www.propagandacritic.com/
articles/intro.ipa.html>.

Ellis, David. *Becoming a Master Student.* Boston: Houghton Mifflin
Company, 1997.

Emilio, George A. *Promoting Critical Thinking in Professional Military
Education.* A research report submitted to the Air Command
and Staff College, Air University, Maxwell Air Force Base,
Alabama. April 2000.

Fisch, Karl, Scott McLeod, and Jeff Brenman. *Did You Know, Video
version 4.0.* New York City: Media Convergence Forum.
October 2009.

Gore, Al. *The Assault on Reason.* New York: The Penguin Press,
2007.

Haskins, Greg R. "A Practical Guide to Critical Thinking." *The Skeptic's Dictionary,* August 15, 2006. <http://www.skepdic. com/essays/haskins.pdf>.

Howley, Kevin. "The War & Peace Report: Democracy Now! and Peace Journalism." A presentation delivered at the Conference on Media War and Conflict Resolution. Bowling Green, Ohio, September 17–19, 2008.

Jamieson, Kathleen Hall. *unSpun:_Finding Facts in a World of Disinformation.* New York: Random House, 2007.

Kandel, Eric. "Illusions." <http://www.alcohol-drug.com/opt_illus. html>

Kaplan, Sheila. *The Powers That Be Lobbying; One Special Interest The Press Doesn't Cover: Itself.* 1988. Republished on-line by The Free Library. <http://www.thefreelibrary.com/

The powers that be lobbying one special interest the press doesn't...-a06860640>. Lewis, Harry. *Academe and the Decline of News Media.* Interview with The Chronicle of Higher Education. November 15, 2009 < *http://chronicle.com/article/Academethe-Decline-of/49120>*

Mander, Alfred E. *Clearer Thinking.* New York: Philosophical Library, 1947.

Messina, James J. *Overview of Critical Thinking.* November 19, 2009. <http://www.livestrong.com/article/14710-overview-of-critical-thinking/>.

Naisbitt, John. *Mind Set*. New York: Collins Business, An Imprint of HarperCollins Publishers, 2006.

Neumann, Robert. "American Democracy At Risk." *Phi Delta Kappan, The Education Policy Magazine of Phi Delta Kappa International*. 2008   <http://www.pdkintl.org/kappan/k_v89/k0801neu.htm>

Paul, Richard, and Linda Elder. *The Miniature Guide to Critical Thinking: Concepts and Tools*. Dillon Beach, CA: Foundation for Critical Thinking Press, 2008.

Pegis, Jessica. *5 Ways to Develop Your Child's Critical Thinking Skills*. GOOGOL Learning. <http://www.googolpower.com/content/articles/5-Ways-to-Develop-Your-Childs-Critical-Thinking-Skills>.

Pew Research Center's Project for Excellence in Journalism. *The State of the News Media—2009*.   <http://www.stateofthemedia.org/2009/narrative_overview_majortrends.php?media=1>

Polmar, Jay. *Think Right; Faith in Beliefs. Intuitive Meaning*. <http://intuitivemeaning.com/2010/02/faith-in-beliefs/>.

Public Broadcasting Service Policy. *Public Broadcasting Service Editorial Standards and Policies*. June 14, 2005. <http://www.pbs.org/aboutpbs/aboutpbs_standards.html>.

Rowland, Willard D., Jr., "Public Broadcasting in the United States." *Encyclopedia of Communication and Information*, 2002: 28.

Rutledge, Pamela. "The Coevolution of Society and Media." *Psychology Today Blog,* October 25, 2009. <http://www. psychologytoday.com/blog/positively-media/200910/the-coevolution-society-and-media>

Sageman, Marc. *Understanding Terror Networks.* Philadelphia: University of Pennsylvania Press, 2004.

Schafersman, Steven D. *An Introduction to Critical Thinking.* January 1991. <http://www.freeinquiry.com/critical-thinking. html>.

Singer, Margaret Thaler. *Thought Reform Exists: Organized, Programmatic Influence.* F.A.C.T.net, 2010. <http://www. factnet.org/Thought_Reform_Exists.htm>.

Stein, Dustin. "Professors Teach What To Think, Not How To Think." *The Daily Campus,* Nov. 18, 2003. <http://www. dailycampus.com/2.7438/professors-teach-what-to-think-not-how-to-think-1.1069956>

Sunstein, Cass R. *On Rumors.* New York: Farrar, Straus and Giroux, 2009.

Sunstein, Cass. *Republic.com 2.0.* Princeton, NJ: Princeton University Press, 2007.

Tietjen, Marjorie. "Defending Ourselves Against Mind Control." Wellsphere, June 26, 2009. <http://stanford.wellsphere. com/lyme-disease-article/defending-ourselves-against-mind-control/733675>.

Walsh, D., and R. Paul. *The Goal of Critical Thinking: From Educational Ideal to Educational Reality.* Washington, DC: American Federation of Teachers, 1998.

Wang, Sam, and Sandra Aamodt. *Welcome to Your Brain: Why You Lose Your Car Keys but Never Forget How to Drive and Other Puzzles of Everyday Life.* New York: Bloomsbury, 2008.

www.eyes-and-vision.com. "A 'Psychological' Optical Illusion." *Vision Health Resources.* <http://www.eyes-and-vision.com/influence-of-culture-on-visual-perception.html>.

www.ingramcontent.com/pod-product-compliance
Lightning Source LLC
Chambersburg PA
CBHW030341290526
45785CB00004B/1551